DNA CLONING/SEQUENCING WORKSHOP
A Short Course

ELLIS HORWOOD SERIES IN BIOCHEMISTRY AND BIOTECHNOLOGY

Series Editor: Dr ALAN WISEMAN, Senior Lecturer in the Division of Biochemistry, University of Surrey, Guildford

* *In preparation*

DNA CLONING/ SEQUENCING WORKSHOP
A Short Course

KEITH FIRMAN B.Sc., Ph.D.
Senior Lecturer
Biophysics Laboratories, Portsmouth Polytechnic

ELLIS HORWOOD

NEW YORK LONDON TORONTO SYDNEY TOKYO SINGAPORE

First published in 1991 by
ELLIS HORWOOD LIMITED
Market Cross House, Cooper Street,
Chichester, West Sussex, PO19 1EB, England

A division of
Simon & Schuster International Group
A Paramount Communications Company

Printed and bound in Great Britain
by Bookcraft (Bath) Limited, Midsomer Norton, Avon

British Library Cataloguing in Publication Data

K. Firman
DNA Cloning/Sequencing Workshop: a short course
CIP catalogue record for this book is available from the British Library
ISBN 0–13–218041–3 (Library Edn.)
ISBN 0–13–218033–2 (Student Edn.)

Library of Congress Cataloging-in-Publication Data available

Acknowledgements

Portsmouth Polytechnic, Biophysics Laboratories are grateful to the following companies for their generous donation of materials for this course.

- Amersham International plc, Lincoln Place, Green End, Aylesbury, Buckinghamshire who supplied the radiochemicals and Sanger sequencing kit.

- Gibco BRL, P.O. Box 35, Trident House, Renfrew Road, Paisley, PA3 4EF for the restriction enzymes, IPTG, X-Gal and other DNA modifying enzymes.

- Anachem, Anachem House, 20 Charles Street, Luton, Bedfordshire, LU2 0EB for the Gilson pipetman P20s and P200s and the yellow tips.

- Pal Europe Ltd, Europe House, Portsmouth for the Biodyne nylon membrane used for Southern blotting.

- Macherey-Nagel GmbH & Co. KG, P.O. Box 101352, D-5160 Duren, West Germany - UK agents: CamLab, for the Nucleobond AX plasmid purification cartridges.

- Boehringer Mannheim UK, (Diagnostics and Biochemicals) Limited, Bell Lane, Lewes, East Sussex, BN7 1LG for the DIG non-radioactive labelling kit.

- United States Biochemical Corporation, PO Box 22400, Cleveland, Ohio 44122, USA - UK agents: Cambridge Bioscience, 42 Devonshire road, Cambridge, CB1 2BL for the Sequenase sequencing system.

Table of Contents

CHAPTER 1

INTRODUCTION

INTRODUCTION

This course is intended as a practical introduction to molecular cloning and its uses. The M13 filamentous coliphage is used as cloning vector and this then allows direct DNA sequence analysis of the cloned DNA fragments. The first experiment is to prepare DNA fragments suitable for cloning. However this experiment has been designed such that sufficient data is obtained to assemble a simple restriction map of the starting material, a recombinant plasmid containing a *Xenopus laevis* ribosomal DNA insert. In parallel to the main cloning and sequencing experiments, a Southern Blotting experiment is included. The data from this experiment allows improvements to the simple restriction fragment map and gives further information about the genomic organisation of the ribosomal DNA.

Procedures used here may in some cases have been slightly modified from those used in the laboratory. This is necessary in order to complete the course in the 4/5 days allotted. However the procedures used still give a good outline of those in common use in research laboratories.

Isolation of plasmid DNA

The isolation of DNA from bacterial cultures is fundamental to genetic engineering. Generally one is interested in isolating either chromosomal DNA (to use in a cloning experiment) or in plasmid DNA (which may be a cloning vector, or may be a recombinant plasmid that has already been prepared). In this brief introduction to DNA cloning you will be given the opportunity to compare three different techniques used to isolate plasmid DNA.

Plasmids are extrachromosomal DNA that exist (usually) as circular super-coiled molecules. They vary in size enormously from as little as a few kilobases up to several hundred kilobases. They often carry an antibiotic resistance gene or some other marker (e.g. heavy metal ion resistance) that is beneficial to the host bacterium. This also allows easy maintenance of the plasmid within the bacterium by growth on selective media.

Cloning vectors are often based on naturally occurring plasmids that have been manipulated to improve selection and to produce a limited number of restriction enzyme sites for DNA cloning.

In the Birnboim & Doly (1979) technique, plasmid DNA is prepared by lysis of bacteria with a solution of sodium dodecyl sulphate (SDS) and NaOH; after treatment with lysozyme to remove the outer cell wall. SDS denatures bacterial proteins and destroys the lipid membrane. NaOH denatures chromosomal and plasmid DNA. The mixture is neutralised with potassium acetate, causing the plasmid DNA to reanneal rapidly. Most of the chromosomal DNA and bacterial proteins precipitate - as does the SDS, which complexes with potassium - and are removed by centrifugation.

Nucleic acids are then concentrated by ethanol precipitation. If ultra-pure plasmid DNA is required then ultracentrifugation in CsCl gradients is usually employed.

The rapid boiling method (Holmes & Quigley, 1981) depends upon the boiling for the denaturation of the chromosomal DNA. The nucleic acids can then be separated from the remaining cell debris by precipitation with propan-2-ol. However, in this course you will use an anion exchange column to purify the DNA. These columns can elute different types of DNA (plasmid, bacteriophage and genomic) at different salt (KCl) concentrations

The last technique used uses differential precipitation of the plasmid DNA with polyethylene glycol (PEG) 6000 followed by isopycnic centrifugation in CsCl gradients. The CsCl gradient will separate supercoiled plasmid DNA molecules from other DNA because of a density difference in the presence of ethidium bromide. Ethidium bromide is an interchelating agent that can "slot" between the bases of the DNA molecule. This results in a lowering of the density of the DNA molecule. However, supercoiled DNA can only allow a limited amount of the interchelating molecule to bind before torsional strain prevents further binding. This results in the supercoiled DNA being more dense than the other DNA species.

Sanger 'dideoxy sequencing'

The Sanger technique for the analysis of DNA sequence undoubtedly represents one of the most significant advances in molecular biology (Williams *et al.*, 1986). The technique is both beutifully simple and extreemly rapid. Its ease of use is, however, closely linked to the development of bacteriophage M13 cloning vectors (Messing & Vieira, 1982); with their ability to rapidly produce large quantities of easily purified single-stranded DNA and to the development of high resolution denaturing polyacrylamide gel electrophoresis.

The technique is dependent upon the ability of the *Escherichia coli* DNA polymerase I enzyme to utilise analogues of deoxyribonucleotides as substrates for DNA sysnthesis. The normal sysnthesis of DNA proceeds in a 5'-3' direction linking the 3'-hydroxyl group of the newly synthesised chain to the 5'phosphate group of the incoming deoxyribonucleotides (Fig. 1.1). This reaction requires only the DNA polymerase I enzyme, Mg^{2+} and the four dNTPs. The analogues used for Sanger sequencing lack the 3'-hydroxyl group required for chain elongation. These analogues are dideoxyribonucleotides (ddNTPs) (Fig. 1.2). When a dideoxyribonucleotide is incorporated into a growing DNA chain then synthesis will stop at that point.

In the Sanger sequencing technique four separate reactions are carried out simultaneously - one for each of the dideoxy bases. Each tube also contains all four dNTPs (one of which is radioactively labelled), Mg^{2+} and the DNA

9

Figure 1.1

Dideoxy Chain Termination

Given a single-stranded template, a complementary primer with a 3'-hydroxyl group and all four deoxynucleoside triphosphates, DNA Polymerase I will catalyse chain extension from the primer, thus synthesising a complementary strand to the template.

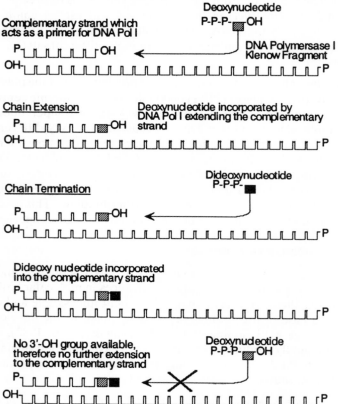

Deoxynucleotide
P-P-P-▨-OH

Complementary strand which acts as a primer for DNA Pol I

DNA Polymersase I
Klenow Fragment

Chain Extension

Deoxynucleotide incorporated by DNA Pol I extending the complementary strand

Chain Termination

Dideoxynucleotide
P-P-P-■

Dideoxy nucleotide incorporated into the complementary strand

No 3'-OH group available, therefore no further extension to the complementary strand

Deoxynucleotide
P-P-P-▨-OH

polymerase I enzyme. Thus the G^o tube contains ddGTP, dGTP, dCTP, dTTP, ^{35}S-dATP, Mg^{2+} and the DNA polymerase I enzyme. DNA synthesis results in random incorporation of the ddGTP and all newly synthesised DNA will end with a ddGTP. By using the correct ratios of dGTP to ddGTP the lengths of newly synthesised DNA can be made totally random from 25bp to a few thousand base-pairs. The four reaction mixes are loaded side by side on the polyacrylamide gel.

In practice a fragment of the *E. coli* DNA polymerase I enzyme is used (the Klenow fragment) this fragment lacks the 5'-3' exonuclease activity of DNA polymerase I and cannot degrade the primer used to initiate DNA synthesis. This is particularly important so as to maintain a fixed 5'-end for the

Figure 1.2

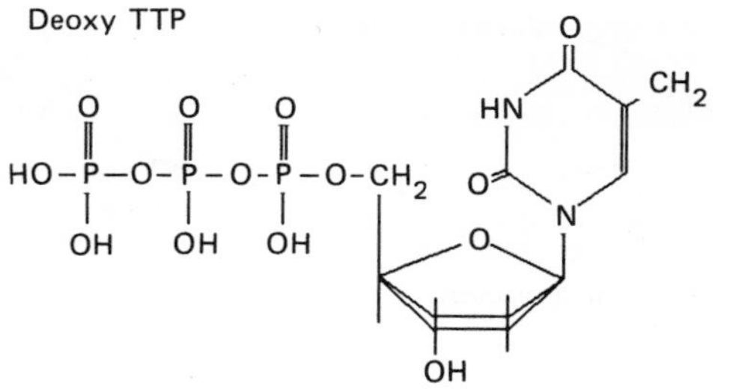

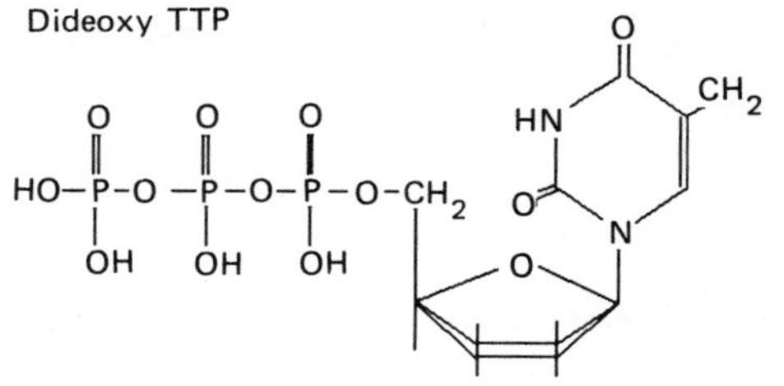

Deoxy TTP

Dideoxy TTP

growing DNA strands. Size separation on high resolution denaturing polyacrylamide gels is from a known point and will guarantee that fragments differing by only one base in length are in fact due to consecutive bases in the sequence (Fig. 1.3). Since one of the dNTPs (usually dATP) is radioactively labelled the fragments on the gel can easily be visualised by autoradiography. The DNA sequence is easily read direct from the autoradiograph by moving up the gel from track to track as the bands appear.

M13 Cloning

The Sanger 'dideoxy sequencing method', although offering a rapid and simple protocol of DNA sequencing, originally required time-consuming methods for producing the essential single-stranded template. These limitations have now been overcome by using a range of M13 vectors prepared by Messing.

M13 is a single-stranded filamentous phage. Its life cycle (figure 1.4) can be exploited for the preparation of pure single-stranded template. The phage enters a suitable host cell (*E. coli* F') by way of the F pilus. On entering the cell, the virus is stripped of its protein coat. The single-stranded viral DNA is then converted to a double-stranded replicative form (RF). This stage is followed by DNA replication to give 100 or more progeny RF molecules from which new single-stranded viral DNA is synthesised. This is then packaged into viral coat proteins and extruded from the host without cell lysis, thus completing the infectious cycle.

11

Figure 1.3

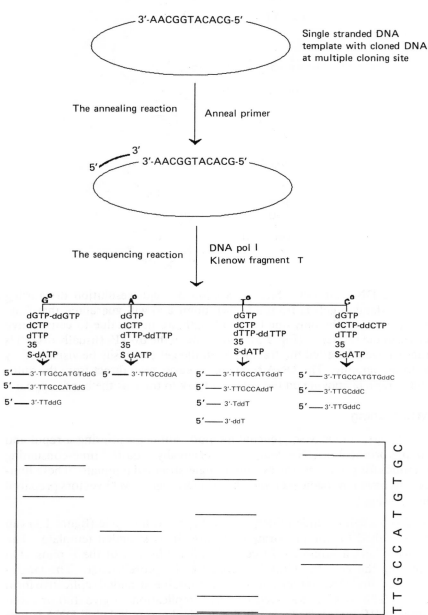

Sanger sequencing

In this way, some 200 phage particles are produced per cell, per generation. The phage can be harvested, free of contaminating material from the host cells, by polyethylene glycol (PEG) precipitation of the culture superna-

Figure 1.4

M13 life cycle

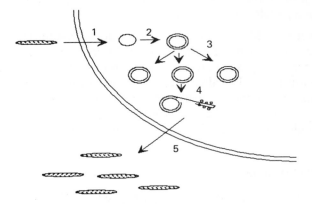

1. Phage enters the bacterial host cell via pilus
2. Single-stranded DNA is converted to double-stranded RF
   (replicative form)
3. Replication of DNA to give progeny RF (100 + per cell)
4. Single strand DNA synthesis by means of the rolling ball method
5. Single-stranded DNA is processed and packaged into mature
   virus and extruded from the cell (200 + per generation)

tant. The single-stranded DNA can then be stripped of its viral protein cost by treatment with phenol. After ethanol precipitation this simple procedure will yield sufficiently pure single-stranded DNA for Sanger 'dideoxy sequencing'.

Insertion of 'foreign DNA'

The M13 life cycle can also be exploited for the preparation of recombinant DNA by using the double-stranded RF DNA as a cloning vector. Fragments of foreign DNA can be inserted into a suitable restriction enzyme site. M13 RF DNA carrying such a double-stranded insert can be introduced into a suitable competent host cell by a transformation step. The resultant phage growth will lead to production of the hybrid molecule in both double-stranded (RF) and single-stranded (mature virus) forms, thus offering both an amplification step and a means of producing the insert DNA in single-stranded form. (The double-stranded hybrid form is also useful for subcloning and similar kinds of manipulation).

Large fragments of duplex DNA can be inserted into an M13 vector because, as it is a filamentous phage, there are no structural restraints on the size of DNA molecule that can be packaged. Molecules with large inserts of DNA simply result in longer phage filaments. 40 Kb inserts are reported to have been cloned into M13; however, in practice, inserts of above 5 Kb tend to be unstable. The size of insert ideally suited for cloning into M13 (300-900bp) is considerably less even than this, since the amount of sequence information that can be determined from a sequencing reaction is limited to 250-400 nucleotides.

The Messing series of M13 strains has been developed to fulfill this role of cloning vector. They have been altered at the intergenic region (position 5489-6005) where it is possible to insert foreign DNA without disrupting essential viral functions. The alterations at this insertion site are such that it now contains several unique restriction enzyme sites. Thus the circular RF molecule can be cut at the insertion site with a suitable restriction enzyme and a piece of foreign DNA can be integrated within it using T4 DNA ligase.

Ligation of DNA molecules

This is the joining of neighbouring 3'-hydroxyl and 5'-phosphate ends of a DNA molecule to produce a phosphodiester bond. The reaction is catalysed by DNA ligases. There are two commonly available DNA ligases:

a) *E. coli* DNA ligase.

This enzyme requires NAD as a co-factor and can catalyse the joining of either nicked DNA or the complementary sequences produced by restriction endonuclease digestion.

b) T4 DNA ligase.

This enzyme requires ATP but has a wider substrate specificity. As well as the types of ligation carried out by *E. coli* DNA ligase, T4 DNA ligase can also catalyse blunt-ended ligation and the joining of double stranded DNA-RNA hybrids. The ability to perform blunt-end ligation is what makes T4 DNA ligase particularly important in DNA cloning.

The main factors affecting ligation are ATP concentration and DNA Concentration. ATP concentration affects the ability of T4 DNA ligase to ligate the above substrates. Increasing the ATP concentration inhibits blunt-ended ligation and DNA-RNA hybrid formation. This enables the sequential ligation of protruding- and blunt-ended fragments in the same reaction simply by altering ATP conc.

When a fragment of DNA ligates it has a choice between inter- and intra-molecular ligation (Revie *et al.*, 1988). The preference is dependent upon both the length of the fragment and the concentration of the fragment. For a constant length then circularisation increases with decreasing concentration. It is possible to describe a concentration of a substrate where the rate of formation of the inter-molecular ligation is equal to the rate of formation of the intra-molecular ligation:

$$j \text{ (g/l)} = 51.1 \times M_r^{-1/2}$$

DNA concentrations smaller than j will drive towards circularisation. In the bimolecular reaction represented by the production of a typical recombinant plasmid the required reaction is favoured by an equimolar concentration of cloning vector and "foreign" DNA. It is, therefore, necessary to calculate the "j-value" for both vector and "foreign" DNA molecules and then to arrange that both molecules be in equimolar concentration but with both the r concentrations above the respective "j-values":

e.g.

pBR322 $Mr = 2.6 \times 10^6$ For 1000bp fragments $M_r = 3 \times 10^6$

$$j = 51.1/(2.6 \times 10^6)^{-1/2} \qquad j = 51.1/(0.6 \times 10^6)^{-1/2}$$

$$= 31.7 \mu g/ml \qquad\qquad = 65.6 \mu g/ml$$

For an equimolar ratio:

$$[\text{foreign DNA}] = \frac{Mr \text{ foreign}}{Mr \text{ vector}} \times [\text{vector DNA}]$$

$$= \frac{0.6 \times 10^6 \times 31.7}{2.6 \times 10^6}$$

$$= 7.32 \mu g/ml$$

This concentration for the "foreign" DNA is less than the "j-value", therefore, circularisation of the foreign DNA will occur. Since to prevent circularisation the foreign DNA must be at a concentration $= 65.6 \mu g/ml$ this value can also be used in the above calculation for equimolar concentrations to give the concentration of vector required for an equimolar reaction:

ie.

$$[\text{vector DNA}] = \frac{M_r \text{ vector}}{M_r \text{ foreign}} \times [\text{foreign DNA}]$$

$$= \frac{2.6 \times 10^6 \times 65.6}{0.6 \times 10^6}$$

$$= 284.3 \mu g/ml$$

This value is well above the j-value for the vector. Therefore, by mixing vector DNA at a concentration of 284.3μg/ml and "foreign" DNA at a concentration of 65.6μg/ml neither molecule will circularise and the preferential reaction should be inter-molecular ligation. Generally transformation of *E. coli* requires only 50μg of DNA in total so only small volumes of the above DNA mix are required.

Selection of recombinants

Although M13 does not lyse its host cell, the growth of the cell is retarded as a consequence of supporting phage growth. Thus, when plated out, those cells infected with M13 will show up as areas of slower growth, which look like turbid plaques on a lawn of uninfected cells. Transformants can therefore be easily distinguished from non-transformed cells. In addition, the Messing M13 strains have been modified to allow the visual discrimination of recombinants from non-recombinants. Under appropriate conditions the M13 vector will grown up to form a blue plaque, but this process is disrupted by the introduction of a fragment of foreign DNA into the M13 insertion site. As a result recombinant phages (those carrying passenger DNA) will give colourless or 'white' plaques.

This discrimination is based on the presence or absence of the enzyme β-galactosidase. *E. coli* host cells infected with any of Messing's M13 vector strains will, in the presence of the Lac operon inducer IPTG (isopropyl-β-D-thiogalactopyranoside), produce a functional β-galactosidase. Such cells will hydrolyse the substrate X-gal (5-bromo-4-chloro-3-indolyl-β-galactoside) to give a blue dye (bromochloroindole). Insertion of foreign DNA into an appropriate site in the DNA vector interferes with the production of ß-galactosidase in infected cells. Thus recombinant M13 gives colourless plaques on an *E. coli* lawn, in contrast to the blue plaques given by the intact vector. The colourless recombinant plaques can be picked out and grown quickly to give a single-stranded template for Sanger 'dideoxy sequencing'.

The M13 vectors mp8, mp9, mp10 and mp11 have several unique restriction enzyme sites in the insertion region. These sites in mp8 and mp9 and in mp10 and mp11, are mirror images of each other (figure 1.5). The advantage of this is that DNA cut with two different restriction enzymes (e.g. *Eco*RI and *Pst*I) can be cloned into, for example, mp8 and mp9 in opposite orientations (figure 1.6). Thus by using mp8 and mp9 or mp10 and mp11 dual vectors, it is possible to sequence both strands from opposite ends of the insert. As discussed earlier the amount of sequence information that

16

Figure 1.5

Multiple Cloning Sites in M13 Messing Vectors

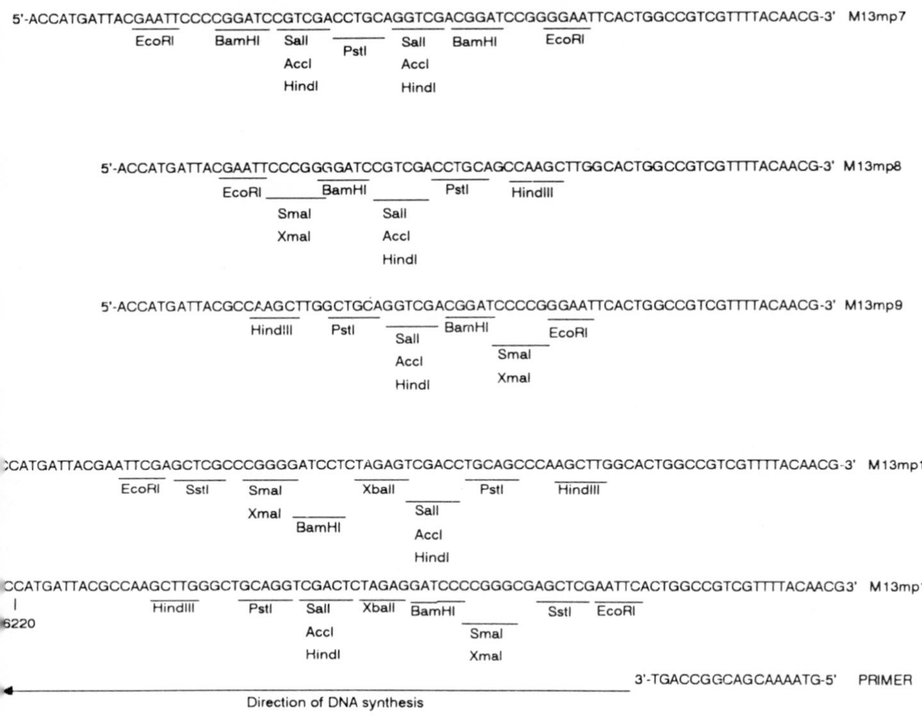

5'-ACCATGATTACGAATTCCCCGGATCCGTCGACCTGCAGGTCGACGGATCCGGGGAATTCACTGGCCGTCGTTTTACAACG-3' M13mp7

EcoRI	BamHI	SalI		SalI	BamHI	EcoRI
		AccI	Pstl	AccI		
		HindI		HindI		

5'-ACCATGATTACGAATTCCCGGGGATCCGTCGACCTGCAGCCAAGCTTGGCACTGGCCGTCGTTTTACAACG-3' M13mp8

EcoRI		BamHI		Pstl	HindIII
	SmaI		SalI		
	XmaI		AccI		
			HindI		

5'-ACCATGATTACGCCAAGCTTGGCTGCAGGTCGACGGATCCCCGGGAATTCACTGGCCGTCGTTTTACAACG-3' M13mp9

HindIII	Pstl	SalI	BamHI	EcoRI
		AccI	SmaI	
		HindI	XmaI	

CCATGATTACGAATTCGAGCTCGCCCGGGGATCCTCTAGAGTCGACCTGCAGCCCAAGCTTGGCACTGGCCGTCGTTTTACAACG-3' M13mp10

EcoRI	SstI	SmaI	XbaII	Pstl	HindIII
		XmaI	SalI		
		BamHI	AccI		
			HindI		

CCATGATTACGCCAAGCTTGGGCTGCAGGTCGACTCTAGAGGATCCCCGGGCGAGCTCGAATTCACTGGCCGTCGTTTTACAACG3' M13mp11

6220

HindIII	Pstl	SalI	XbaII	BamHI	SstI	EcoRI
		AccI		SmaI		
		HindI		XmaI		

3'-TGACCGGCAGCAAAATG-5' PRIMER

◄——————————————————————————————————
Direction of DNA synthesis

can be determined from a sequencing reaction is limited to 250-400 nucleotides. However, when using either mp8 and mp9 or mp10 and mp11 in unison, longer sequences can be determined by sequencing a large insert 250-400 nucleotides from both ends. It is also possible to confirm sequencing data by determining the sequence of both strands of a shorter fragment.

The 'universal' primer

Before the development of M13 cloning as a system for generating single-stranded template, it was necessary to prepare a separate primer for each fragment of DNA to be sequenced by the Sanger 'dideoxy method'. It is now possible to use a primer complementary to region of M13 DNA which flanks the insertion site (figure 1.5). Thus the Klenow fragment of DNA polymerase I will extend the primer in a 5' to 3' direction, passing immediately through the cloning site and so will give the sequence of any DNA fragment inserted at this position. Such a primer is often referred to as 'universal' since only one primer is required for any fragment of DNA that is cloned into the insertion site.

17

Figure 1.6

Sequencing in opposite directions using M13mp8 and M13mp9

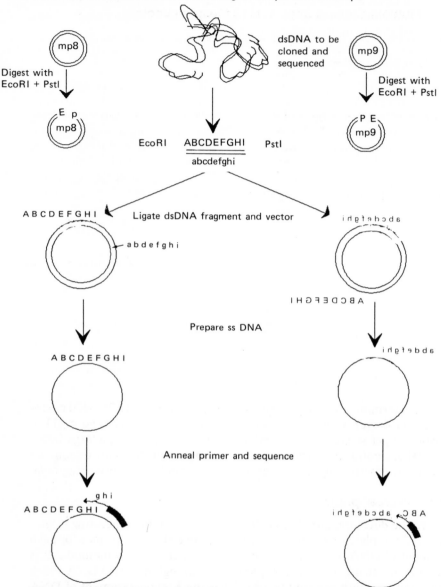

The universal primer was originally a restriction fragment with a sequence complementary to a region immediately downstream from the insertion site. This fragment (usually about 26 bp long) has now been superseded by a shorter synthetic primer of 15 or 17 bases which anneals more efficiently and exhibits no self-annealing.

Thus the development of M13 cloning systems has overcome the two earlier disadvantages of the 'dideoxy sequencing method'. Large amount of single-stranded DNA can now be readily prepared and a universal primer can be used. This combination of M13 cloning and 'dideoxy sequencing' offers a rapid and simple method of sequencing DNA.

Unidirectional deletions

While the use of "complementary pairs" of M13 cloning vectors allows both strands of a DNA sequence to be read with large inserts it is still extreemly difficult to sequence beyond 450bp. One can prepare new sequencing primers to regions of the DNA insert that have been sequenced but this is time consuming and expensive. However, the multiple cloning sites of M13mp8 and M13mp9 have allowed the development of an altenative strategy for sequencing long regions of DNA (Ozkaynak & Putney, 1987). This technique involves random, overlapping deletions of the region that

Figure 1.7

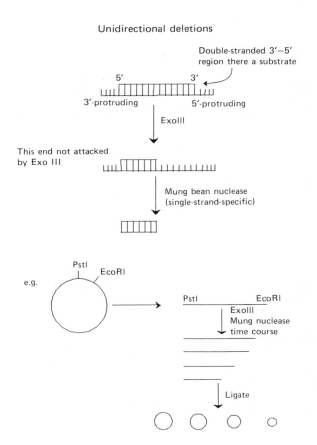

Unidirectional deletions

has already been sequenced. The basis of the technique is that the enzyme exonuclease III of *E. coli* is a **double-strand-specific** 3'-5' exonuclease and will degrade one strand of a 5'-protruding end of a DNA molecule but not a 3'-protruding end. By cutting the multiple cloning site, at the end where the universal primer site lies, with an enzyme that produces a 3'-protruding end the primer site is "protected" from exonuclease III digestion. The end of the multiple cloning site adjacent to the insert is then cut with a restriction enzyme that produces a 5'-protruding end. Digestion by exonuclease III over a short time course followed by digestion with the single strand specific Mung Bean nuclease results in deletion of the insert DNA. The blunt ends produced can be re-ligated using T4 DNA ligase and then the remaining insert DNA sequenced (see figure 1.7).

Southern Blotting

This is a procedure by which the restriction pattern of chromosomal DNA fragments can be analysed. Chromosomal DNA is restricted with various enzymes, separated by gel electrophoresis and a copy of the gel separation

Figure 1.8

Southern blotting

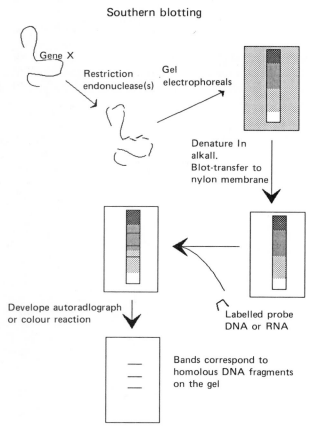

20

made by transferring the DNA, after denaturation, to a nitrocellulose or similar membrane.

The membrane is later saturated with various macromolecules to avoid further DNA binding to it. It is then bathed in a solution of a denatured and radioactively labelled DNA such that this DNA can hybridise to homologous sequences bound on the membrane (Southern, 1975). When excess radioactivity has been removed the membrane is autoradiographed, (see figure 1.8).

Random Priming Labelling of DNA

To prepare a DNA probe for Southern Blotting it is necessary to label the DNA molecule to be used. This labelling usually involves the incorporation of a radioactive nucleotide into the DNA. The two most commonly used techniques available for radioactively labelling DNA are random priming and nick translation.

In the random priming technique a mixture of synthesised hexanucleotides (covering all the possible combinations of bases) are annealed to a single-stranded template in a manner analogous to that used in Sanger sequencing. The single-stranded template can be produced either by the use of M13 cloning vectors or by boiling and then rapidly cooling a linearised double-stranded DNA molecule. The annealed primers are then used to prime second strand synthesis using the Klenow fragment of $E.$ $coli$ DNA polymerase I and a mixture of all four nucleotides. If one of the four nucleotides has a ^{32}P at the α-position then the newly synthesised strand will be radioactive (see figure 1.9).

Figure 1.9   Random primer labelling

Random heaxanucleotide primers

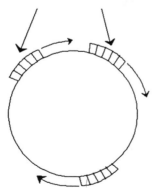

Klenow, dNTPs
and labelled
nucleotide

Nick Translation of DNA

E. coli DNA polymerase I adds nucleotide residues to the 3'-hydroxyl terminus that is created when one strand of a double-stranded DNA molecule is nicked. The nick in a double-stranded DNA molecule can be created using the enzyme DNase I. In addition, the polymerase, by virtue of its 5' to 3' exonucleolytic activity, can remove nucleotides from the 5' side of the nick. The elimination of nucleotides from the 5' side and the sequential addition of nucleotides to the 3' side results in movement of the nick (nick translation) along the DNA (see figure 1.10). By replacing the pre-existing nucleotides with highly radioactive nucleotides, it is possible to prepare ^{32}P-labelled DNA with a specific activity well in excess of 10^8 cpm/µg.

Figure 1.10

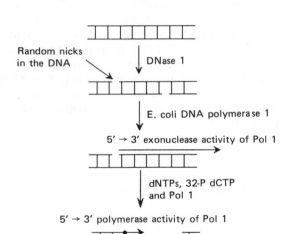

Nick translation of DNA

Non-radioactive Labelling of DNA

Non-radioactive labelling of DNA has a number of advantages over conventional labelling using ^{32}P. Firstly, the non-radioactive label is much safer to use (hence its use in this practical) it also removes the problems of disposal and containment. Secondly, the labelled DNA is stable for at least 12 months and can easily be stored for repeated use.

The Boehringer DIG^{TM} system is based upon an alkaline phosphatase enzyme linked to an antibody to digoxigenin hapten (see figure 1.10)

22

(digoxigenin is a steroid which occurs naturally only in digatalis plants - the antibody therefore shows no cross reaction with other steroids). The

Figure 1.11

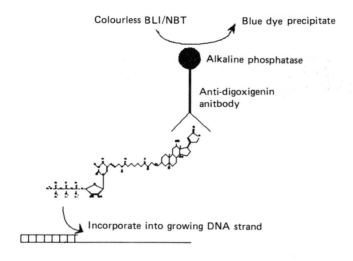

Non-radioactive labelling of DNA

Colourless BLI/NBT → Blue dye precipitate

Alkaline phosphatase

Anti-digoxigenin anitbody

Incorporate into growing DNA strand

Structure of Digoxigenin-11-UTP

digoxigenin hapten is covalently attached to dUTP (see figure 1.11) which can be incorporated into a newly synthesised DNA strand either by nick translation of random priming. The colour reaction involves the production of a blue dye precipitate from 5-Bromo-4-chloro-3-indolylphosphate/Nitroblue tetrazolium chloride(BCIP/NBT). This precipitate remains bound to the nylon or nitrocellulose membrane and can be easily photographed or photocopied.

23

References:

BIRNBOIM H.C. & DOLY J. (1979) Nucleic Acids Res. *7*, 1513-1523

HOLMES D.S. & QUIGLEY M. (1981) Anal. Biochem. *114*, 193-197

MESSING J. & VIEIRA J. (1982) Gene *19*, 269-276

OZKAYNAK F. & PUTNEY S.D. (1987) Biotechniques *5*, 770-773

REVIE, D., SMITH, D.W. & YEE, T.W. (1988) Nucleic Acids Res. *16*, 10301-10321

SOUTHERN, E.M. (1975) J. Mol. Biol. *98*, 503-518

WILLIAMS S.A. SLATKO B.E. MORAN L.S. & DESIMONE S.M. (1986) Biotechniques *4*, 138-147

CHAPTER 2

SUMMARY OF CLONING PROCEDURES

Summary of Practical Procedures

Day (1)

(a)   Restriction of Plasmid DNA.

(b)   Ligation of DNA to cloning vector.

(c)   Restriction of chromosomal DNA for Southern Blot.

(d)   Small scale plasmid isolation.

(e)   Start cells for Day 2 transformation and large scale plasmid DNA isolation.

Day (2)

(a)   Prepare competent cells.

(b)   Transform *E. coli* with *in vitro* recombined DNA.

(c)   Southern Blot.

(d)   Large scale plasmid DNA isolation.

(e)   Start cells for Day 3.

Day (3)

(a)   Prepare ssDNA from recombinant Clone.

(b)   Preparation of non-radioactive DNA probe

(c)   Pre-hybridisation and Hybridisation of Southern Blot.

(d)   Purification and analysis of prepared DNA.

Day (4)

(a)   Sequence Analysis of Recombinant Clone.

(b)   Wash and develope colour reaction for the Southern Blot.

(c)   Unidirectional deletions of plasmid DNA.

(d)   Demonstration of sequencing gel pouring.

(e)   Autoradiography of sequencing gels

Day (5)

(a)   Development of autoradiographs.

(b)   Analysis of DNA sequences by computer.

(c)   Discussion of results.

CHAPTER 3

PRACTICAL PROCEDURES

Restriction of plasmid DNA

<u>Day 1</u>

(a) Complete digests.

    ● **Working on ice:**

1.   Take 1µl of the ribosomal plasmid pXl108c solution 1µg/µl). Transfer to a 1.5 ml microcentrifuge (micro-reaction or Eppendorf) tube. Add 1µl of 10x concentrated restriction buffer (10 x CORE Buffer) and 8µl sterile distilled water, (dH$_2$O). Mix well by flicking tube with fingers and then centrifuge briefly.

2.   Repeat step (1) until four duplicate tubes are set up. Label tubes:

> pXl108c
> pXl108c/*Eco*RI
> pXl108c/*Pst*I
> and pX108c/*Eco*RI + *Pst*I.

3.   To the appropriate tubes add 1µl (3 units) of restriction enzyme *Eco*RI or *Pst*I or both. Mix by flicking tube then briefly centrifuge.

4.   Transfer tubes, excepting that marked pX108c, to the 37^{0}C water bath. Leave for one hour or a little more.

5.   While waiting for the restriction reactions prepare a 1% analytical agarose gel as in Appendix 1.

6.   When restriction reactions are complete transfer tubes to a 65^{0}C water bath to "kill" enzymes. Leave for 5 min and finally return to ice.

7.   Pipette 8µl of 1 x concentrated agarose gel loading buffer (1 x L.B.) into 1.5ml tube. Repeat until four tubes are prepared. Label these as in (2) above.

8.   Pipette 2µl from each restriction reaction into the appropriate tube containing gel loading buffer.

Generally one tends to load appprox. 100ng of DNA on an agarose gel.

However, when small fragment (< 1Kb) are to be visualised then it may be necessary to increase this amount.

When mixing the contents of a microfuge tube always spin the tube briefly to bring the contents to the bottom of the tube.

Loading buffer contains glycerol as well as bromophenol blue and is used to increase the density of the solution so as to facilitate loading into the wells or pockets of the gel.

Day 1

Notes

(b) Partial digestion of DNA.

- **This technique is both useful for resolving restriction fragment maps - it can give information regarding adjacent fragment and for DNA cloning where the fragment to be cloned contains a restriction site for the enzyme being used to clone with.**

1. Take $2\mu l$ of the ribosomal plasmid pXl108c solution $1\mu g/\mu l$). Transfer to a 1.5 ml microcentrifuge (micro-reaction or Eppendorf) tube - label the tube **TUBE 0**. Add $1\mu l$ of 10 x concentrated restriction buffer (10 x CORE Buffer) and $8\mu l$ sterile distilled water, (dH$_2$O). Mix well by flicking tube with fingers and then centrifuge briefly.

2. Number 8 microfuge tubes through 1-8. Pipette $8\mu l$ of sterile water into each tube and place them in a 65°C water bath.

3. Add 1 unit of *Pst*I to **tube 0** and immeadiately start a stop clock.

4. At 30 sec intervals remove $1\mu l$ of solution from **tube 0** and place into one of the tubes numbered 1-8. Seal the tube and leave at 65°C for 10 mins.

    - **This will stop the *Pst*I digestion and, therfore, give a time course for digestion.**

5. Add $1\mu l$ of 5 x loading buffer to each DNA sample.

    - **Using a pipette load these solutions and the complete digest solutions in known order into the pockets of the agarose gel. Reserve the first pocket for $10\mu l$ of molecular weight marker solution.**

6. Run gel as described in Appendix I.

    - **While the gel is running begin the chromosomal DNA digestion, Day 1.**

    - **When gel is complete and a photographic record has been obtained continue with the DNA isolation as follows.**

When using lambda markers remember to heat the marker at 65°C to melt the cohesive (*cos*) end of the lambda molecule.

Rapid isolation of plasmid DNA

Day 1.

(a) The alkaline lysis method:

1. From a 2.0ml overnight culture of a strain of *E. coli* containing a recombinant plasmid centrifuge 1.5ml of this culture for 1 min in a microcentrifuge.

2. Remove the growth medium with a pasteur pipette.

3. Resuspend the pellet by vortexing in 100μl of an ice-cold solution of:

> 50mM glucose
> 10mM EDTA
> 25mM Tris-HCl (pH8.0)
> 4mg/ml lysozyme.

4. Leave at room temperature for 5 min.

5. Add 200μl of a freshly prepared, ice-cold solution of:

> 0.2M NaOH
> 1% SDS.

6. Mix the tube by inversion (2-3 times).

   ● **DO NOT VORTEX.**

7. Store on ice for 5 min.

8. Add 150μl of an ice-cold solution of:

> 60ml of 5M potassium acetate
> 11.5ml acetic acid
> 28.5ml H2O.

9. Gently vortex for 10 sec. Store on ice for 5 min.

10. Centrifuge in a microcentrifuge for 5 min.

11. Transfer the supernatant to a fresh 1.5ml Eppendorf tube.

    ● **CARE!! WEAR GLOVES WHEN HANDLING PHENOL - BURNS!!**

Plasmid containing bacteria grown in L-broth + selective antibiotc (Ampicillin @ 150μg/ml).

Denatures the chromosomal DNA.

Precipitates the chromosomal DNA before it can re-nature.

12. Add an equal volume of phenol/chloroform. Mix by vortexing.

13. Centrifuge in a microcentrifuge for 2 min. Transfer the supernatant to a fresh tube.

14. Add two volumes of cold (-70°C) ethanol, mix by vortexing and leave at room temperature for 30 min.

15. Centrifuge for 5 min in a bench centrifuge.

16. Discard the supernatant.

17. Add 1ml of 70% ethanol and re-centrifuge.

18. Discard the supernatant. Dry the pellet by inverting the tube in a 37°C incubator.

19. Add 50μl of TE buffer containing DNase-free RNase (20μg/ml). Vortex briefly. Incubate at 37°C for 30 min.

20. Repeat steps 12-18. Resuspend the pellet in 50μl of TE buffer.

● **Store the DNA solution at -20°C for future analysis**

The RNase solution must be heated to 95°C for 5 min to destroy DNase activity.

(b) Medium Scale Rapid Boiling Method

1. Spin down a 1.5 ml overnight culture of the plasmid containing bacteria. Resuspend the pellet by careful pipetting in 100μl of STET buffer.

2. Add 5μl of fresh lysozyme soln.

3. Place the microcentrifuge tube in a dry heating block at 100°C for 1 min..

4. Spin the contents of the tube for 10 mins.

5. Equilibrate a Nucleobond AX cartridge with 0.8ml of buffer N2 (stand the cartridge in a 14ml centrifuge tube). Carefully remove the cleared lysate from the microcentrifuge tube and pass it through the cartridge .

6. Wash the cartridge with 3 x 0.7ml of buffer N3.

7. Elute the plasmid DNA into a clean 14ml centrifuge tube with 0.8ml of buffer N5.

8.

Precipitate the DNA with 0.8 volumes of cold -20oC propan-2-ol.

9. Spin down the DNA/RNA with a 5 min spin. Dry the pellet and resuspend in 100μl of water.

    ● **Store DNA as before**

    ● **Continue with the ligation of DNA fragments**

Ligation of DNA fragments

Day 1

● **Working on ice:**

1. Ligate *Pst*I digested pXl108c with the *Pst*I digested M13 vector provided as follows:

● **Pipette into four 1.5ml microfuge tubes labelled a - d:**

1μl 10 x ligase reaction buffer
2μl 10 mM ATP/50 mM DTT mix
3μl d H_2O

2. To tube a add:

2μl (200 ng) pX1l108c/*Pst*I

To tube b add:

1μl (100ng) pX1l108c/*Pst*I

To tube c add:

0.5μl (50ng) pX1l108c/*Pst*I

To each tube (including d) add:

2μl (20 ng) M13mp18 vector

3. Mix and centrifuge briefly. Add 1μl (0.1 units) of T4 ligase. Again mix and centrifuge briefly. Label the tubes with your name and leave O/N at $\sim 4^0$C or $\sim 14^0$C.

● **Setup an overnight gel of the chromosomal DNA digest**

● **From plate of *E. coli* TG1 pick a single colony and transfer to 10ml of 2 x TY medium. Grow overnight (O/N) shaking at 37^0C.**

● **Use the supplied 1.5ml O/N culture to inoculate 150ml of L-broth containing 150μg/ml ampicillin for the large scale DNA isolation.**

Transformation of E. coli

1. From overnight (O/N) culture of TG1 take 0.2ml and use to inoculate a fresh 10ml of 2 x TY medium. Label this "Competent Cells". Shake at 37^{0}C for 3 hr.

   Grow to mid-log.

2. Inoculate a further 10ml of 2 x TY with only a drop each of TG1 O/N culture. Shake at 37^{0}C. This will provide fresh cells for use at the plating and growing up stages, see below.

   The fresh cells are required to enable the M13 to re-infect health, growing cells.

   - **Use time waiting for cultures in continuing with Southern Blot protocol.**

3. Spin down 6 x 1.5ml of the 3 hr culture of TG1 labelled "Competent Cells" in a microfuge for, 1 min.

   - **You prepare one tube of compotent cells per transformation.**

4. Pour off the media into 10% Chloros and resuspend the cells in 450μl of sterile, ice cold MOPS (pH7.0).

5. Re-spin the cells for 1 min. Resuspend in 450μl of cold MOPS/CaCl2 (pH6.5). Leave on ice for 90 mins.

   - **Start the large scale DNA isolation.**

6. Spin down the cells for 1 min in a microfuge, discard the supernatant and then carefully re-suspend the cells in 150μl of MOPS/CaCl2 (pH6.5). Keep on ice.

7. Take six 1.5ml sterile microfuge tubes. Label the tubes:

   > 1 = "Non-ligated vector"
   > 2 = "Ligated vector DNA"
   > 3 = "pXL108c/M13mp18 ligation"(x3)
   > 4 = "No DNA".

8. To the tube labelled "Non-ligated vector" add 1μl of M13mp18 vector DNA from Day (1). To the tube labelled ligated vector add the contents of the overnight ligation tube d. To the "No DNA" tube add 5μl of sterile water. To the other tubes add 5μl of the overnight ligation reactions (a, b and c respectively).

9. Add the competent cells to all of the tubes. Mix well and leave on ice for 30 minutes.

● **Continue with the large scale DNA isolation.**

10.   Heat shock both tubes at 65^0C for 30 sec, then return im-
       meadiately to ice.

11.   Mix:

           240μl of 100 mM IPTG
           240μl of 2% X-gal
           1200μl of fresh TG1 culture in a 4ml tube
           on ice.

12.   Add 270μl of this mix to each 1.5ml tube.

13.   Label six L-agar plates:

           1 = "Non-ligated vector"
           2 = "Ligated vector DNA"
           3 = "pXL108c/M13mp18 ligation" (x3)
           4 = "No DNA".
           and your name, on the bottom of the plate.

14.   Remove each 1.5ml tube from ice. Add the contents to 3ml
       of molten H-top agar (kept at 42^0C in a 4ml test tube). Mix
       by rolling tube and immediately pour onto surface of the
       appropriate H-plate.

15.   Leave both plates to set at R/T. Invert and incubate O/N at
       37^0C.

● **From plate of** *E. coli* **TG1 pick a single colony and transfer to
10ml of L broth. Grow overnight (O/N) shaking at 37^0C.**

Large scale isolation of plasmid DNA:

<u>Day 2</u>

Isopycnic centrifugation:

1.   Spin down the cells in a 250ml bucket at 5000 r.p.m. for 15 min in a Sorval RC5-B or equivalent.

2.   Pour of the supernatant and carefully resuspend the cells in 6.6ml of 25% sucrose solution. Avoid any pellet by repeatedly pipetting.

   ● **CARRY OUT THE FOLLOWING OPERATIONS ON ICE!**

3.   Add 1.34ml of FRESH lysozyme solution (5mg/ml).

4.   Leave for 5 min. on ice.

5.   Add 2.6ml of 0.25M EDTA.

6.   Leave for 5 min. on ice.

7.   After 5 min. remove a 1ml sample, add 1ml of Triton solution and gently mix. If the cells lyse - indicated by the solution becoming thick and viscous proceed to stage 10. If this does not occur leave the cells another 5 min.

8.   Add 10.5ml of Triton to each of the complete buckets carefully down the side of the bucket so that the triton floats on the surface.

   ● **N.B. add 1ml less to the bucket from which the sample was removed.**

9.   Carefully mix the two layers by repeated pipetting through an inverted 10ml pipette until lysis occurs.

10.   Centrifuge at 10,000 r.p.m. for 30 mins in a Sorval RC5-B or equivalent.

11.   Carefully remove the cleared lysate from the viscous pellet - it's a bit like separating egg yolks from whites.

12.   Add a 1/2 volume of 30% PEG soln. Leave at room temperature for 30 mins.

13. Spin-down the precipitated nucleic acids (etc) in a 50ml centrifuge tube at 10,000 r.p.m. for 10 mins. in a Sorvall RC5-B.

14. Resuspend the pellet in 11ml of TE buffer - it may require a vortex mixer to help resuspend the pellet.

15. Add 11.38g of CsCl, 0.22ml of TE and 0.35ml of Ethidium bromide. Fill a quick seal centrifuge tube and centrifuge in a Beckman 70-1Ti rotor at 55,000 for 16 hrs.

   ● **Load the large scale DNA preparations into the ultracentrifuge.**

Day (3)

   ● **It is now necessary to remove the cccDNA from the CsCl and ethidium bromide.**

   ● **Avoid mixing the contents of the centrifuge tube.**

1. Carefully remove the top of the Quick seal Tube with a pair of sharp scisors or wire cutters.

2. Remove a small amount of the CsCl solution from the top of the tube.

3. With a Pasteur pipette in a 2ml Pi-pump carefully remove the lower DNA band into a large test tube.

4. Add an equal volume of butan-2-ol to the CsCl/DNA solution. Shake vigourously and then allow the tube contents to settle.

   ● **The ethidium bromide will partition into the upper butanol layer.**

5. Remove and discard the upper layer.

6. Repeat stages 4 & 5 until no colour remains in the CsCl/DNA solution.

7. Pipette the CsCl/DNA solution into a dialysis bag and dialyse for 1 hour against two 1 litre changes of TE buffer.

   ● **This DNA and that from the small scale preps. will be analysed by gel electrophoresis on day 4.**

   ● **The DNA can now be cut with a restriction enzyme such as *Eco*RI to check its purity.**

Dialysis can be replaced by dilution in TE buffer followed by ethanol precipitation.

Single-strand DNA isolation

Day (3)

1. Remove plates from 37^0C incubation. On the "Ligated DNA" plate you should observe many small round clear areas in the lawn of bacteria. These are the M13 plaques. Some plaques will show a blue tint, these contain wild type M13, others will show no blue colouration, these contain recombinants.

    ● **To prepare single stranded viral DNA from one recombinant plaque:**

2. Dispense a 15μl aliquot of O/N TG1 culture into a sterile 30ml culture tube containing 1.5ml 2xTY medium.

3. Inoculate tube with a colourless plaque using sterile wooden cocktail sticks (toothpicks should not be used as these are sometimes treated with anti-microbial agents).

4. Shake the tubes at 37^0C for >5 hours. Transfer to 1.5ml microcentrifuge tubes.

    ● **Start the preparation of a non-radioactively labelled DNA probe.**

    ● **Use this time to purify the large scale DNA preparation from the CsCl.**

Vigorous shaking is required for growth periods of 5-7 hours. Overnight growth can be used if very slow shaking is used (<100rpm).

5. Centrifuge for 5 minutes. (This, and subsequent steps, are performed at room temperature). Pour supernatant into a fresh tube, being very <u>careful not to pick up any cells</u> - do not take all of supernatant. Retain cell pellet.

This preciptates the phage.

6. Re-centrifuge supernatant as before, to ensure that all cells are removed.

7. Add 1200μl of supernatant to 200μl PEG/NaCl. Shake, then leave to stand for 15 minutes.

8. Centrifuge for 5 minutes. Discard supernatant.

Removes the protein coat of the phage.

9. Centrifuge for 2 minutes. Carefully remove all remaining traces of PEG with a "drawn-out" Pasteur pipette. Wipe off any traces of PEG on mouth of tube with tissue. A viral pellet should be observable at this stage.

10. Add 100μl TE buffer to the viral pellet, vortex well. Add 50μl of phenol saturated with TE buffer. Vortex 30 seconds.

11.   Stand tubes for 1 minutes at ambient temperature. Vortex 30 seconds. Repeat this process twice.

12.   Centrifuge for 3 minutes.

13.   Remove and transfer upper (aqueous) layer to a fresh microcentrifuge tube.

14.   Add one-half volume of CHCl3, vortex for 30 sec and carefully remove the top (aqueous) layer to a fresh microcentrifuge tube.

15.   Add 10μl 3M Na-acetate. Add 250μl of ethanol and mix by inverting tube.

16.   Leave overnight at-20^{0}C to precipitate the DNA.

Ethanol precipitation can be for a minimum of 2 hrs at -20°C or for 30 mins at -70°C. Always add 2.5 volumes of COLD ethanol.

Day 4

17.   Centrifuge the overnight precipitation of viral DNA for 10 min.

18.   Pour off supernatant, add 1ml of cold (-20^{0}C) ethanol and centrifuge for 5 min.

Take great care not to loose the DNA pellet - it is not always easily visible!

19.   Pour off ethanol and leave inverted on a tissue to drain dry.

20.   When completely dry re-dissolve DNA in 50μl of TE by vortexing liquid for about 1 min. Store on ice.

   ● **Analyse all prepared DNA by gel electrophoresis (use *EcoRI* to cleave the double stranded DNA and run these digests against uncut plasmid).**

Practical Procedures for DNA sequencing

Day (4)

(a) Klenow sequencing.

● **To perform a sequence analysis of the inserted DNA the primer is first annealed to the ssDNA, the template.**

1. Pipette into a 1.5ml microcentrifuge tube:

    > 5μl of the template DNA
    > 1μl of the M13 primer
    > 1.5μl of Klenow reaction buffer
    > 2.5μl of dH$_2$O

● **Make sure the tubes are properly sealed.**

2. Briefly centrifuge tube, flick to mix contents, then briefly centrifuge again.

3. Incubate tube in a water bath or oven at 65°C for 20 min and then allow to cool slowly to room temperature.

● **Working at R/T carry out sequencing reactions as follows:**

4. After the annealing reaction is complete briefly centrifuge tube.

5. Take care to wear gloves and dispose of used tips and tubes in radioactive waste container as you are now dealing with radioactive sulphur (^{35}S). Add 1μl ($10\,\mu$Ci) of (^{35}S)dATP to annealed primer/template mix.

6. Add 1μl of Klenow fragment (of DNA polymerase I from *E. coli*) at 1 unit/μl, to primer template mix. Mix and briefly centrifuge.

7. Place four 1.5ml tubes labelled A, C, G and T into centrifuge.

8. To each of these tubes add 2.5μl of the annealed primer/template + enzyme mix.

9. To each tube carefully add 2μl of the relevant dNTP/ddNTP mixes. Add the 2μl drop such that it does not contact the 2.5μl drop of primer/template + enzyme.

i.e.

> tube A + 2μl A^0/ddATP
> tube C + 2μl C^0/ddCTP
> tube G + 2μl G^0/ddGTP
> tube T + 2μl T^0/ddTTP

10. Centrifuge briefly to start the reaction. Incubate for 20 min at 37^{0}C.

Notes

11. Exactly 20 mins later add 2μl of dNTP chase mix exactly adding the dNTP/ddNTP mixes. Allow reaction to proceed for a further 20 min at 37^{0}C.

(b) Sequenase™ sequencing.

1. Pipette into a 1.5ml microcentrifuge tube:

   > 5μl template DNA (500ng)
   > 3μl of H$_2$O
   > 1μl of Sequencing buffer
   > 1μl of primer

2. Anneal by heating for 2min. at 65°C then cool slowly to room temperature.

3. While cooling, label four 1.5ml microcentrifuge tubes G, A, T and C. Dispense 2.5μl of the supplied termination mix into the appropriate tube.

4. Add 8μl of H$_2$O to 2μl of labeling mix.

5. Pre-warm the 4 termination tubes from step 3 in a 37°C water bath.

6. Dilute 1.5μl of sequenase™ enzyme with 10.5μl of TE buffer.

7. Prepare the labeling reaction:

   > 10μl of ANNEALING MIXTURE (From step 1)
   > 1μl of 0.1M DTT
   > 2μl diluted LABELING MIX (From step 4)
   > 0.5μl ^{35}S dATP
   > 2μl diluted Sequenase™ enzyme (From step 6)

   > Mix the tubes and incubate at room temperature for 5min.

44

8.   Transfer 3.5µl of labeling reaction to each of the four pre-
      viously prepared termination tubes (G,A,T, and C), mix and
      continue incubation at 37°C for 5 min.

9.   Stop the reactions by adding 4µl of formamide dye.

      ● **During Day (4) you will be shown how the polyacrylamide/urea
        sequencing gels are poured and set up. When gels are ready
        bring your samples to be loaded onto a sequencing gel.**

17.   The samples will first be denatured by heating to ~ 95°C for
      3 min then 2µl loaded into a gel slot. After running the gels
      hot at 50 to 60 watt for several hours, the DNA is fixed in the
      gel in a 2 1 bath of 10% acetic acid, 10% methanol for 20 min,
      dried onto a sheet of filter paper under vacuum at ~ 80°C.
      This is necessary in order to detect the weak radiation from
      the ^{35}S during the subsequent autoradiography.

Unidirectional deletion of cloned DNA.

1. Pipette 5μl (5μg) of an *Eco*RI/*Pst*I double digest of pTZ18R DNA into a 1.5ml microcentrifuge tube.

> The double digest produces one 5'- and one 3'-protruding end on the DNA molecule.

2. Add 6μl of 10 x ExoIII buffer and 59μl of dH$_2$O.

3. Number 12 microcentrifuge tubes and pipette 7.5μl of MUNG MIX into each tube and store on ice:

   20μl 10 x Mung buffer
   50-60 units of Mung Bean Nuclease
   Make to 200μl with dH$_2$O

4. Warm the DNA solution to 37°C in a water bath and add 250 units of ExoIII. Immeadiately start a timer.

5. Remove 2.5 μl samples, at 1 minute intervals, into the cold MUNG MIX.

6. Quickly spin the tubes contianing DNA + MUNG MIX in a microfuge and then incubate at room temperature for 20 mins.

7. Add 1μl of STOP BUFFER and heat to 65°C for 10 mins to inactivate the Mung Bean Nuclease.

   ● **Pour a 0.8% agarose gel.**

8. Analyse the extent of the deletion by running 2μl samples of the digested DNA against the *Eco*RI/*Pst*I double digest of pTZ18R DNA and lambda marker.

Restriction enzyme digestion of chromosomal DNA

Day (1)

• **Working on ice:**

1.   Take 9μl ($\sim 3\mu$g) of *Xenopus laevis* chromosomal DNA and transfer to 1.5ml tube. Add 1μl of 10x restriction buffer (10 x CORE Buffer). Mix thoroughly (the DNA is extremely viscous) and centrifuge briefly.

2.   Repeat (1) and label the two tubes *Eco*RI and *Pst*I.

3.   To the appropriate tubes add 2μl (6 units) of *Eco*RI or *Pst*I enzyme. Again mix thoroughly and centrifuge briefly.

4.   Transfer tubes to 37^0C water bath for one hour or more.

• **Continue with the small scale DNA isolation.**

5.   After more than an hour at 37^0C transfer tubes to a 65^0C water bath for 5 min to "kill" enzymes. Finally return to ice.

6.   At a convenient moment, prepare a 0.8% agarose submarine gel as in Appendix I.

7.   To *Eco*RI and *Pst*I labelled tubes add 3μl of 5 x concentrated agarose gel loading buffer (5 x L.B.).

8.   Load 10μl of molecular weight marker and all solution from the *Eco*RI and *Pst*I tubes into adjacent slots of the agarose gel, noting order.

9.   Run gel not as in Appendix I but at 15V O/N.

> *Notes*
>
> A vast excess of restriction enzyme is used to ensure complete digestion.

Practical Procedures for Southern Blotting

Day (2)

Notes

1. Make a photographic record of the O/N gel electrophoresis (with a ruler alongside gel).

2. Trim away unused portions of the gel with a scalpel to leave 0.5cm on either side of the used tracks.

   ● **Measure the size of the gel.**

   The time taken to accomplish transfer of the DNA from the agarose to the filter can be reduced markedly by nicking the DNA within the gel. Soak the gel for 5 min in 0.25M HCl.

3. Denature DNA by transferring to a large volume ($\sim$200ml) of 1.5 M NaCl and 0.5 M NaOH. Leave 30 min at R/T, agitating intermittently.

4. Neutralise gel by soaking for a further 30 min in 1 M Tris-HCL pH 8, 1.5 M NaCl. Agitate intermittently.

5. Using a scalpel carefully cut a piece of nylon membrane to the same size as the trimmed gel.

   Nitrocellulose can be used instead of nylon but is more difficult to handle.

6. Using scissors or a guillotine, cut 10cm high stack of paper towels to about the same size as trimmed gel. Cut two pieces of 3 MM paper to the same size as the gel. Cut one piece larger than the gel as well.

   Several sheets of 3MM can be used instead of the foam.

7. When gel is neutralised, thoroughly soak a piece of plastic foam with 10 x S.S.C. (standard saline citrate) in a plastic dish.

   ● **See Figure   - how to assemble a Southern Blot.**

8. Place the large sheet of 3 MM paper on the foam making sure no air bubbles are trapped.

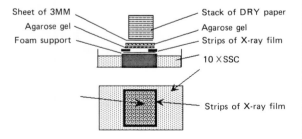

Sheet of 3MM — Stack of DRY paper
Agarose gel — Agarose gel
Foam support — Strips of X-ray film
— 10 X SSC
— Strips of X-ray film

9. Place trimmed gel centrally onto 3 MM covered foam. Make sure no bubbles are trapped. Note its orientation.

10. Float nylon on a distilled water surface until it is thoroughly wetted, (colour will change to light grey). Pour off distilled water and replace with 2 x S.S.C. submerging nitrocellulose.

11. Place strips of waterproof material under the edge of the gel (approx 2mm overlap of gel) all around the gel. This prevents liquid contact between source of 10 x S.S.C. and dry paper towels except through gel, see Fig 1.

We routinely use strips of X-ray film from discarded sequencing autoradiographs.

12. Using gloves transfer wetted nylon onto gel, excluding all bubbles.

13. Again excluding bubbles overlay nitrocellulose with the last two sheets of 3 MM paper soaked in 2 x S.S.C. Finally overlay these with the complete stack of dry cut down paper towels and use a gel electrophoresis plate to weigh the whole stack down.

14. Top up the plastic tray with 10 x S.S.C. until liquid level lies 1cm below the top of the foam.

- **As the S.S.C. is soaked upwards through the gel, the DNA will move slowly to the nylon where it will be trapped. Leave until the 10 x SSC is approximately three-quarters of the way up the dry stack.**

- **Continue with the transformation of *E. coli*.**

A ball point pen will write on a nylon membrane.

Day 3

15. Carefully remove paper towel stack and 3 MM papers and mark positions of gel slots on nitrocellulose membrane using a pencil or ball point pen. Also mark orientation of membrane relative to gel so that slots can be later identified.

16. Carefully remove nylon using gloves and forceps. Transfer to 2 x S.S.C. for 5 min.

Alternatively nylon filters can be fixed by 3 min exposure to UV light.

17. Drain excess S.S.C. from nitrocellulose and lay on filter paper until dry.

18. Place nylon between the leaves of a dry 3 MM paper folded in half and bake at 80^{0}C for 2 hours to fix DNA.

Non-radioactive labelling of DNA.

Day 3

1. Place 1μl (1μg) of linearised (by restriction enzyme digestion) pXl108c DNA in a microcentrifuge tube. Denature the DNA by heating at 95°C for 10min and cooling quickly in an ice/NaCl bath.

2. Add 2μl of hexanucleotide mix (vial 5), 2μl of dNTP labeling mix (vial 6) and 14μl of sterile H2O (Final volume should be 19μl).

3. Add 1μl of Klenow enzyme (2 units).

4. Centrifuge briefly and incubate for 60 mins at 37°C.

    ● **Return to purifying the large scale preparation of DNA.**

5. Add 2μl 0.2M EDTA solution (pH8.0)

6. Precipitate with:

    > 2.5μl 4M LiCl
    > 75μl pre-chilled ethanol (-20°C)

7. Mix well and leave at -70°C for 30min.

    ● **Prehybridise the filter from the Southern Blot.**

8. Centrifuge in a microcentrifuge for 10min., wash with 40μl of cold ethanol (70% v/v). Re-centrifuge and dry.

9. Dissolve for 30 min in 50μl 10mM Tris-HCl, 1mM EDTA (pH8.0)

10. Remove a 2μl sample add 50μl of TE and place in a boiling water bath for 5 min and then transfer the entire 50μl immeadiately into 2.5ml of the pre-warmed (to 68°C) hybridisation solution.

    ● **Move immediately to the hybridisation of probe and DNA.**

This is random priming - nick translation can also be used.

NaCl, LiCl or sodium acetate can all be used to help precipitate DNA in ethanol.

Prehybridisation of the filter.

Day 3

1. Place the filter between two pieces of plastic film and seal three edges.

2. Add 20ml of prehybridisation solution into the bag, dispel any air bubbles and then seal the bag.

3. Incubate at 68°C for at least one-hour.

4. Open the prehybridisation bag and empty out the prehybridisation solution but do not let the filter dry before hybridisation.

> The prehybridisation solution is used to block all the nylon membrane that does not have ssDNA bound to it. The blocking agent used is dried milk solids.

Hybridisation of the filter with the non-radioactive probe.

Day 3

1. Transfer the probe, in 2.5ml of pre-warmed hybridisation solution, into the bag containing the filter.

2. Incubate at 68°C for at least 6 hrs (O/N).

Day 4

3. Remove the filter from the bag and wash in a small tray with 50ml of 2 x SSC, 0.1% SDS at 68°C for 5 min.

4. Repeat step 3 with fresh SSC/SDS.

5. Wash with 50 ml of 0.1 x SSC, 0.1% SDS for 15 min at 68°C.

6. Repeat step 5 with fresh SSC/SDS.

Overnight hybridisation is convenient here.

This is a high stringency wash - only extreemly homologous sequences will hybridise.

Detecting hybridised DNA.

Day 4

1. Wash the filter for 1 min with BUFFER 1:

    100mM Tris-HCl pH7.5
    150mM NaCl

2. Incubate the filter for 30 min with 50ml of BUFFER 2:

    1% Blocking Reagent in BUFFER 1

3. Repeat step 1.

4. Pipette $4\mu l$ of antibody-conjugate (vial 8) into 20ml of BUFF-ER 1. Incubate the filter for 30 min with this solution.

5. Wash the unbound antibody-conjugate off the filter with two 15 min washes with 100ml of BUFFER 1.

6. Wash the filter with 20ml of BUFFER 3:

    100mM Tris-HCl pH9.5
    100mM NaCl
    50mM $MgCl_2$

7. Prepare the COLOUR SOLUTION:

    $45\mu l$ NBT-solution (vial 9)
    $35\mu l$ X-phosphate-solution (vial 10)
    10 ml BUFFER 3

8. Place the filter in a plastic bag, add the COLOUR SOLU-TION and seal the end of the bag. Leave the filter face down O/N in a dark cupboard for the colour to develope.

Day 5

9. Pour off the COLOUR SOLUTION and wash the filter for 5 min with 50ml of TE buffer.

    ● **The results can be recorded by photocopying.**

Analysis of Results

<u>Day (5)</u>

- This day is set aside for the development of autoradiograph and the analysis of results.

- Analysis of the restriction data from Day (1) will be limited to constructing possible restriction maps of the plasmid pX1108c.

- Analysis of the M13 clones is made in these practicals only via the DNA sequence. If you wish you may prepare an agarose gel and analyse some of the viral DNA directly. Also by a very simple procedure the double strand R.F. form of the virus contained in the cell pellet of Day (3) can be analysed on this gel. This will give an idea of the size of the DNA insert and hence which fragment you have cloned.

- Details of how to read the DNA sequence are contained in figure 1.3.

- A computer generated DOTPLOT of the cloned *X. laevis* rRNA spacer against the published sequence can be prepared.

- If you wish the computer can also be used to carry out analysis of other DNA and protein sequences.

CHAPTER 4

ADDITIONAL PRACTICAL
PROCEDURES

Preparation of M13 bacteriophage

M13 bacteriophage are secreted from an *E. coli* host as single-stranded DNA surrounded by a protein coat. Large titres of such bacteriophage can easily be obtained from the supernatant of a suitable overnight culture. The starting point for the growth of such a culture is usually a single plaque (i.e. from an wild type M13 transformation) such plaques can be picked into 100μl of Hershey's buffer and stored at -20°C.

(a) Preparation of phage.

1. Dilute one drop (approx. 20μl) of an overnight culture of JM109 or TG1 into 20ml of fresh L-broth.

   Store phage at -20°C

2. Pick a fresh M13 plaque into 100μl of Hershey's buffer in a microcentrifuge tube. Use 50μl of this phage suspension to inoculate the dilute culture of JM109.

3. Grow overnight with vigorous shaking.

4. Spin down the bacteria at 10,000 r.p.m. and carefully remove the supernatant (M13 bacteriophage).

5. Re-spin a small sample of the bacteriophage in a microcentrifuge to ascertain that NO bacteria remain in the phage suspension. Store the bacteriophage suspension at 4°C (keeps for approx. 3 months).

(b) Ascertaining phage titre.

6. Melt 6ml of soft L-agar and dispense 2 x 3ml into small (4ml) test tubes in a heating block at 48°C.

   0.1ml of phage into 0.9ml of Hershey's buffer per serial dilution.

7. Mix 100μl of a log phase culture of JM109 (or TG1) with the soft agar and spread the contents of each tube over the surface of a L-agar plate. Allow the top agar to set.

8. Prepare a dilution series of the bacteriophage stock from 10^0 through 10^{-7} in Hershey's buffer. Carefully place 10μl spots of each dilution on the surface of the plates (a max. of 5 spots per plate can easily be accommodated.

   Usually $> 10^{10}$ pfu/ml.

9. Incubate the plates at 37°C overnight. Count the plaques at an appropriate dilution and use this to calculate the titre as plaque forming units (pfu)/ml.

M13 Replicative-form (RF) DNA Isolation

The technique described here for the isolation of double-stranded M13 DNA is based upon **chloramphenicol amplification.** This is where the antibiotic chloramphenicol is used to "turn-off" the bacterial host protein synthesis and, therefore, chromosomal replication but allows the plasmid DNA replication to continue since this is independent of *de-novo* protein synthesis.

1. Inoculate 150ml of L-broth with 0.5ml of an O/N culture of JM109. Grow the bacteria to mid-log (A_{450nm} = 1.0) with shaking.

2. Inoculate the bacterial culture with M13mp18 bacteriophage at a multiplicity of infection (MOI) of 10 (Approx. 1.5×10^{11} bacteriophage per culture).

3. Grow the culture for 90 mins with vigorous shaking.

4. Add 1.5ml of chloramphenicol solution (7mg/ml in 50% ethanol)

5. Grow for one hour. Harvest the cells by centrifugation at 10,000 r.p.m. for 10min in a 6 x 250ml rotor.

    ● The dsDNA can now be isolated following any of the techniques described for plasmid DNA isolation.

Prepared as described above.

Phagemid single-stranded DNA isolation:

Phagemids are small plasmid-based cloning vectors that also carry the origin of replication from the single-stranded bacteriophage M13. They are particularly useful in that they combine the ease of plasmid DNA isolation with the ability to sequence using conventional techniques.

1. Inoculate 2ml of L-broth (+ antibiotics and 0.001% B1) with 40µl of an overnight culture of the phagemid containing strain. Grow to mid-log at 37°C with shaking.

2. Add KO7 helper phage to (multiplicity of infection) MOI of 10. Grow for 1 hr with shaking.

3. Mix 40µl of the above culture with 0.5ml of L-broth (+ antibiotics, 0.001%B1 AND KANAMYCIN - 70µg/ml). Grow for 1-5 hrs.

4. Use 20µl of the above culture to inoculate 2ml of L-broth (+ antibiotics, B1 and kanamycin) grow O/N.

    ● **Follow the procedures described in the main text for single strand DNA isolation.**

Prepared as described for M13.

M13 KO7 carries kanamycin-resistance.

Plasmid DNA transformation

● follow steps 1-10 as described in the *transformation of DNA* section of Chapter 3.

11   Spin down the cells in a microcentrifuge tube and resuspend in 1.5ml of L-broth.

12.   Transfer to a culture tube and grow for 30mins at 37°C with shaking.

● This allows the cells time to recover from the CaCl$_2$ and temperature shock.

The M13 transformation has a lawn of JM109 bacteria for phage growth.

13.   Spin down the cells in a microcentrifuge tube again. Resuspend in 100μl of Hershey's buffer and spread the cells onto the surface of a selective agar plate.

● The selective media used will depend upon he cloning vector or plasmid used.

14.   Incubate overnight at 37°C.

Phosphotase treatment of vector DNA

- Alkaline phosphotase is used to remove the 5'-phosphate groups from the ends of vector DNA that has been linearised by one or more restriction endonucleases. This prevents the vector re-ligating and, therefore, increases the yield of recombinant DNA.

1. Place 1µl of vector DNA (1µg/µl) into a microcentrifuge tube.

2. Add 1µl of 10 x CORE buffer and 6µl of sterile water

3. Add 1µl (1 unit) of calf alkaline phosphotase (CAP) and 1µl (10 units of a suitable restriction endonuclease(s).

4. Incubate at 37°C for 1-2 hours (to ensure complete digestion).

   - Check the digestion by diluting 1µl of the DNA 1:10 in sterile water. Add loading buffer and run against 100ng of un-cut DNA on a 1% agarose gel.

5. Phenol extract, chloroform extract and ethanol precipitate the DNA as described in the *preparation of single stranded DNA* section of Chapter 3.

6. Spin-down the DNA pellet for 5 mins in a microcentrifuge. Dry the pellet and resuspend in 7-8µl of sterile water.

   - The effectiveness of the phosphotase can be checked by the ligation and transformation frequency of the phosphotased DNA compared to non-phosphotased vector.

Preparing acrylamide sequencing gels:

(a) Non-gradient gels:

- Clean both sequencing plates thoroughly with soap + water followed by water and finally ethanol. Siliconise (say every 10 gels) the front plate.

1. In a 250ml conical flask weigh 42g of urea. Add 20ml of 40% acrylamide soln. and 10ml of 10 x TBE.

2. Dissolve by gentle heating and stirring, make to 100ml with water.

3. Cool to room temp. add 750µl of 10% AMPS and 40ul of TEMED. Shake thoroughly.

4. Immediately pour between sequencing gel glass plates. Clamp top and bottom and place comb, flat side into gel, in place.

5. Leave at least one hour to set.

6. Assemble sequencing gel into gel box, flush the top of the gel with 1 X TBE. Place 1 x TBE in the bottom chamber. Pre-electroforese for 30 mins at 60 watts, constant power.

7. Immediately flush the top of the gel, insert comb with the teeth 1mm into the gel and load your samples. Run the gel at 60 watts constant power.

(b) Gradient gels:

1. Prepare:

    i. 60ml of:
    1 x TBE gel mix
    360µl 10% AMPS
    30µl TEMED

    ii. 15ml of:
    5 x TBE gel mix
    30µl 10% AMPS
    5µl TEMED

2.   Take 8ml of 1 x TBE gel mix + TEMED into a 25ml pipette. Carefully pipette a further 12ml of 5 x TBE gel mix + TEMED such that an interface occurs.

3.   Allow one or two air bubbles into the pipette to mix the layers.

4.   Carefully pour all the contents of the pipette down one side of the gel plate sandwich.

5.   Refill the pipette from the 1 x TBE gel mix + TEMED and pour gently down the other side of the plates until the interface reaches the centre of the plates. Continue to fill from the other side of the gel.

6.   Clamp the gel and place the comb in place (flat part of comb in gel).

7.   Leave to set.

8.   Assemble gel into gel box, place 500ml of 1 x TBE in the top chamber, flush the top of the gel and insert the comb.

9.   Place 500ml of 2 x TBE in the bottom chamber.

10. Load the samples and run the gel at 60 watts constant power.

(c) Fixing and autoradiography of sequencing gels:

1.   Gently prise apart the gel plates and note which plate the gel has adhered to - this should be the non-siliconised plate.

2.   Place the gel and plate into a large tray and gently cover with 10% acetic acid. Leave for 10 mins.

3.   Remove from the acetic acid and drain excess liquid away.

4.   Gently layer a piece of Whatman 3MM chromatography paper onto the gel. Remove air bubbles by rolling a glass rod gently over the paper.

5.   Lift the paper + gel from the glass plate. Cover the gel with Saran wrap.

6.   Dry on a gel dryer at 80°C for 40 mins under vacuum.

7.   Remove the Saran wrap and in a dark room place a piece of X-ray film next to the gel in a film cassette box. Leave at -70°C for 15 hrs.

8.   Develope the film in X-ray film developer for 5 mins. Fix for 5 mins. Wash and dry.

APPENDIXES

APPENDIXES

Appendix I

Analytical Agarose "Submarine" Gel

● **For a 1% agarose gel:**

1   Weigh 1g agarose onto foil on a top pan balance and transfer to 250 ml conical flask.

2   Add 100ml electrophoresis buffer. Take care to use disposable gloves as this buffer contains ethidium bromide, a known carcinogen.

3   Cover conical flask with foil and bring solution to boil on hot plate or bunsen. Leave boiling for about 1 minute then transfer to 65^{0}C water bath to cool.

4   Seal ends of electrophoresis plate carefully with "Scotch" or masking tape and put slot-former or comb in position.

5   When agarose has cooled to 65^{0}C pour (wearing gloves) onto electrophoresis plate. Spread evenly by tilting plate.

6   After leaving gel for 30 min to set, overlay gel with distilled water and carefully remove slot former. Then remove tape sealing ends of plate.

7   Place gel on plate into the electrophoresis apparatus so that slots or wells are at the negative (black) side. Fill apparatus with electrophoresis buffer until gel itself is just submerged. Replace apparatus cover until gel is required.

To Load and Run Gel:

8   Remove apparatus cover. Carefully pipette 10-20µl of sample into the wells made by the slot former. Take care to eject each sample slowly from the pipette so as to produce the minimum of turbulence.

9   When all solutions have been loaded, replace the lid of the electrophoresis apparatus and connect cables to the power supply. DNA will move towards the positive (red) terminal. Apply 30v to the gel and allow to electrophoresis for 15 min until DNA enters the gel. Then increase the voltage to 100v for about 1 hour or until the blue marker dye, bromophenol blue, is half way down the gel.

10 Make a photographic record of gel analysis by placing the gel on a 300nm U.V. transilluminator and photographing through a 4 x red filter. Ethidium bromide interchelates the DNA bases and gives a strong near U.V. radiation.

- **DO NOT under any circumstances look into the 300nm U.V. source without wearing the face shield provided. Also do not expose hands or face to the U.V. source.**

Appendix II

Lambda DNA Markers

Lambda DNA

*Hind*III

23465 + —————————

9536 ————————

4422 + —————————

2320 ————————
2002 ————————

565 ——————————

125 ——————————

Lambda DNA

*Hind*III + *Eco*RI

————————— 21531 +

—————— 5216
—————— 5078

—————— 4320

——————— 2590 +

—————— 2002
—————— 1934

—————— 1584

—————— 1373

—————— 947

—————— 832

——————— 565

—————— 125

+These lambda fragments heve "sticky ends" and will tend to run as one fragment unless heated to 65°C for 1-2 min before electrophoresis.

Appendix III

Useful hints for handling small volumes:

1   The Gilson Pipetman P20 will accurately measure volumes of between $2\mu l$ and $20\mu l$. It is frequently used to measure $0.5\mu l$ and $1\mu l$, however, it must be remembered that these volumes are not particularly accurate.

2   When measuring these small volumes into an Eppendorf microcentrifuge tube the best means of assuring that the drop is transferred from the tip of the Gilson into the Eppendorf tube is to follow these instructions carefully.

3   Carefully expel the small drop from the tip (drops less than $5\mu l$ hang at the end of the tip).

4   If the Eppendorf tube is empty the touch the drop to the bottom of the tube. The drop will transfer quickly and easily to the bottom of the tube. However, be aware of the possibility of static electricity drawing the drop back to the tip.

5   If there is already liquid in the Eppendorf tube then two choices present themselves.

If there is no reason to worry about the timing of the mixing of the liquids then place the pipette tip directly into the solution in the Eppendorf and expel the contents of the tip. Then carefully pipette the mixture two or three times to mix.

When you want to control the timing of the mixing the carefully pipette the small drops of liquid onto the wall of the Eppendorf tube. The drop will remain in place (provided they are less than $4\mu l$) until the tube is briefly centrifuged.

6   Always operate the pipette slowly to avoid the liquid "flying" up the pipette tip.

7   Check the weight of small volumes frequently and service the pipettes regularly.

8   It is a good idea to use the same pipette throughout multiple operations such as DNA sequencing - this maintains a constant error throughout the reaction volumes.

Appendix IV

List of Main Requirements

Day 1

Requirements	Amount Required	Recipe
1 x loading buffer	50μl	Dilute 5 x loading buffer
5 x loading buffer	50μl	50% glycerol 100mM EDTA pH8.0 0.125% bromophenol blue 0.125% xylene cyanol FF 0.25ml H₂O
10 x ligase buffer	20μl	0.5M Tris-HCl pH7.4 0.1M MgCl₂
10ml pipettes	1 canister (sterile)	
1ml pipettes	1 canister (sterile)	
14ml culture tube	5	
30ml culture tube	1	
37°C water bath	1	
65°C water bath	1	
70% ethanol (@ -20°C)	100ml bottle/class	
Agarose	2g in a bottle	
10 x ATP/DTT	50μl	10mM ATP 50mM DTT

Item	Quantity	Details
Top pan balance	1 per class	
Bunsen	1	
Nucleobond™ AX Column materials	1ml buffer N2 3ml buffer N2 1ml buffer N5	
CORE buffer	10µl	500mM Tris-HCl pH7.5 500mM NaCl 100mM MgCl$_2$ 10mM DTT
Electrophoresis buffer	2 litre	30g glycine 1.2g NaOH 1.5g EDTA 2l H$_2$O 250µl Ethidium bromide solution
O/N culture of C600 (pCP1005)	10ml	L-broth + Amp (150µl/ml
Eppendorf tubes	1 jar	
Ethanol (@ -20°C)	100ml bottle/class	
Ethidium bromide soln.	100µl	10mg/ml
Filler for Pasteurs	1	
Forceps	1	
Gloves	1 box each size per class	
Glucose/EDTA/Lysozyme	500µl	50mM glucose 10mM EDTA 25mM Tris-HCl (pH8.0) 4mg/ml lysozyme.

Horizontal gel apparatus	1
Ice bucket-broth	1
Propanol-2-ol (@-20°C)	100ml bottle/class
Lambda DNA marker	20μl	lambda DNA (30μg/μl) cut with *EcoRI*/*Hind*III and *Hind*III in 1 x loading buffer
L-broth in a 500ml flask + AMP	150ml	10g Tryptone 5g yeast extract 5g NaCl 1g glucose
Gas lighter	1/class
Lysozyme soln.	20μl	5mg/ml (FRESH)
M13mp18/*Pst*I vector	10μl
Masking tape	1 roll/class
Microcentrifuge	1 /3 students
Microwave cooker or bunsen & tripod	1/class
NaOH/SDS	500μl	0.2M NaOH 1% SDS.

Gilson Pipetman P20	1	
Gilson Pipetman P200	1	
Pasteur pipettes	1 box per class	
Phenol/TE mixed 1:1 with CHCl₃.	100ml bottle	
Pi-pumps	1 of each size per student	
Plate of TG1 or JM109	1	
Polaroid camera	1	
Polaroid film	1 pack per four students	
Potassium acetate	500µl	5M
Power pack	1	
pXL108c plasmid DNA	5µl	
BRL restriction enzyme *EcoR*I	20 units	
BRL restriction enzyme *Pst*I	20 units	
Spatulas	1	
Sterile water	100ml bottle	
STET buffer	10 ml bottle	8% sucrose 5% Triton X-100 50mM EDTA 50mM Tris-HCl pH8.0
BRL T4 DNA ligase	10 units	
TE buffer	100ml bottle	10mM Tris-HCl pH 7.5 1mM EDTA

75

Toothpicks	pack of 10
Transilluminator	1 /class
Vortex mixer	1
Weighing boats	2
X. laevis **chromosomal DNA**	10μl
Gilson yellow tips (P20/P200)	1 pack

Day 2

Requirements	Amount Required	Recipe
10 x SSC	2l	1.5M NaCl 0.15M Sodium citrate pH7.0
10ml pipettes	1 canister	
1ml pipettes	1 canister	
25% Sucrose/Tris	10ml	25% sucrose 0.1M Tris-HCl pH8.0
250ml centrifuge bucket	1	
30% PEG soln.	25ml	30% PEG 8000 1.5M NaCl
30ml culture tubes	4	
3MM chromatography paper	1 sheet	
50ml centrifuge tube	1	
50ml measuring cylinder	1	
65°C Water bath	1 per class	
Bunsen	1	
Cling-film	1 roll per class	
CsCl	15g	
Denaturing solution	250ml	1.5M NaCl 0.5M NaOH

Item	Quantity	Notes
EDTA soln.	5ml	0.25M EDTA pH8.0
Eppendorf tubes	1 bottle	
Ethidium bromide soln.	100µl	10mg/ml
Filler for Pasteurs	1	
Foam	1	
Forceps	1	
Gloves	1 box each size per class	
Guillotine	1 per class	
Heating block	1 per class	
High speed centrifuge	1 per class	
Ice bucket	1	
BRL IPTG	250µl	100mM BRL IPTG
L-agar plates	6	
L-broth	100ml bottle	1g Tryptone 0.5g yeast extract 0.5g NaCl 0.1g glucose
Lysozyme soln.	2ml	5mg/ml (FRESH)
Microcentrifuge	1	
MOPS pH7.0	10ml	10mM MOPS pH7.0 10mM RbCl

Item	Quantity	Composition
MOPS/CaCl₂ pH 6.5	10ml	10mM MOPS pH6.5, 10mM RbCl, 50mM CaCl₂
Neutralising solution	250ml	1M Tris-HCl pH5.5, 1.5M NaCl
GILSON PIPETMAN P20	1	
GILSON PIPETMAN P200	1	
Pal Biodyne nylon membrane	1 roll per class	
Paper towels	1 pack	
Pasteur pipettes (long form)	1 box per class	
Plastic dishes (or enamel dishes)	1	
Plate of TG1	1	
Polaroid camera	1 per class	
Polaroid film	1 pack per 4 students	
Quick-seal centrifuge tube	1	
Scalpel	1	
Scissors	1	
Small glass test tubes (4ml capacity)	10	
Strips of X-ray film	4 (15cm x 1 cm)	
TE buffer	1 litre	10mM Tris-HCl pH 7.5, 1mM EDTA

Test tubes (large)	5	
Test tubes (small)	7	
Toothpicks	1 pack of ten	
Top agar	50ml	0.6% agar Autoclave 121°C for 15 min.
Transilluminator	1 per class	
Triton soln.	15ml	1% Triton X-100 0.1M Tris-HCl pH8.0
Vortex mixer	1	
BRL X-GAL	250µl	2% BRL X-GAL in dimethyformamide
Gilson yellow tips (P20/P200)	1 pack	

Day 3

Requirements	Amount Required	Recipe
10ml pipettes	1 canister	
1ml pipettes	1 canister	
30 ml culture tubes	2	
37°C water bath	1 /class	
68°C water bath	1 /class	
95°C water bath	1 /class	
Bag sealer	1 /class	
Bunsen	1	
Butan-2-ol	25ml	
$CHCl_3$	1ml	
Cold 70% ethanol (-20°C)	100ml bottle /class	
Cold ethanol (-20°C)	100ml bottle /class	
CsCl waste bottle	1/class	
Dialysis tubing	15cm length	Pre-treat by boiling in 0.2M EDTA. Store in cold (-20°C) ethanol.

dNTP mix (vial 6 - BOEHRINGER DIG kit)	5µl	
Drawn-out Pasteur pipette	1	
EDTA soln.	5µl	0.2M
Forceps	1	
Gloves	1 box of each size per class	
Hexanucleotide mix (vial 5 - BOEHRINGER DIG kit)	5µl	
Hybridisation mix	3ml (Heat to 65°C for 20-30 min to dissolve blocking agent)	5 x SSC 0.1% Sarkosyl 0.02% SDS 1% Blocking reagent (vial 11 - Boehringer DIG kit)
Ice/salt bath	1	
BRL Klenow fragment of *E. coli* Pol I.	2µl	
LiCl soln.	5µl	4M LiCl

Item	Quantity/Details
Microcentrifuge	1
Microcentrifuge tubes	1 bottle
Na acetate	100μl
3M sodium acetate	
GILSON PIPETMAN P20	1
GILSON PIPETMAN P200	1
Pasteur pipettes	1 box per class
PEG/NaCl	500μl — 20% PEG 6000 / 2.5M NaCl
Phenol/TE	100μl
Pi-pumps (particularly a 2ml)	1 of each size
Plastic lunch box	1
Plastic sheeting	2 sheets 20cm x 5 cm
Pre-hybridisation mix	50ml — 5 x SSC / 0.1% Sarkosyl / 0.02% SDS / 1% Blocking reagent (vial 11 - Boehringer DIG kit)

pXL108c/*Hind*III plasmid DNA	1µg	
Scissors	1	
Solvent waste bottle	1	
Sterile water	100ml bottle	
TE buffer	3l	10mM Tris-HCl pH7.5 1mM EDTA
Test tubes (large)	5	
Toothpicks	1 pack of 10	
Vortex mixer	1	
Wire cutters or scissors	1	
Gilson yellow tips (P20/P200)	1 pack	As supplied by Phamacia

Day 4

Requirements	Amount Required	Recipe
10 x ExoIII buffer	10µl	660mM Tris-HCl pH8.0 6.6mM MgCl$_2$
10 x Mung buffer	100µl	500mM Na acetate pH5.0 300mM NaCl 10mM ZnSO$_4$
10 x TBE	20ml	109g Tris 55g boric acid 9.3g EDTA 1 litre H$_2$O
40% acrylamide	50ml	38g acrylamide 2g Bis acrylamide 100ml H$_2$O
Amersham ^{35}S d-ATP	20µCi	
37°C water bath	1 per class	
65°C water bath	1 per class	
A-mix	5µl	500µl 0.5mM dTTP 500µl 0.5mM dCTP 500µl 0.5mM dGTP 1µl 10mM ddATP 500µl TE buffer
Acetic acid/Methanol	2.5l /class	10% Acetic acid 10% methanol

Agarose	1.0g	
Antibody/conjugate (BOEHRINGER DIG vial 8)	5µl	
BUFFER 1	500ml	100mM Tris-HCl pH7.5 150mM NaCl
BUFFER 2	100ml	1% Blocking agent (Boehringer DIG vial 11) in BUFFER 1
BUFFER 3	100ml	100mM Tris-HCl pH9.5 100mM NaCl 50mM MgCl$_2$
C-mix	4µl	500µl 0.5mM dTTP 25µl 0.5mM dCTP 500µl 0.5mM dGTP 8µl 10mM ddCTP 1000µl TE buffer
Chase mix	10µl	0.25mM dTTP 0.25mM dCTP 0.25mM dGTP 0.25mM dATP in TE buffer
Disposal facilities for radiochemicals	1	
DTT soln.	5µl	100mM DTT
Exonuclease III	250 units	
Formamide dye	50µl	0.1g xylene cyanol FF 0.1g bromophenol blue 2ml 0.5M EDTA 100ml formamide

G-mix   5μl   500μl 0.5mM dTTP
                                 500μl 0.5mM dCTP
                                 25μl 0.5mM dGTP
                                 16μl 10mM ddGTP
                                 1000μl TE buffer

Gel apparatus

Item	Quantity
Gloves	1 box each size per class
Heating block	1 per class
BRL Klenow	2 units
Lunch box	1
Masking tape	1 roll /class
Microcentrifuge	1
Microcentrifuge tubes	1 bottle
Beta-monitor for radioactivity	1
Mung Bean nuclease	50 units
NBT soln. (Boehringer DIG vial 9)	50μl
GILSON PIPETMAN P20	1
GILSON PIPETMAN P200	1
Plastic sheeting	2 pieces 25cm x 10 cm
Power pack	1
Sequenase™	2 units
Sequenase™ labelling mix	5μl
Sequenase™ termination mix	15μl
Sequencing gels	2 /class 42g Urea 20ml 40% acrylamide soln. 10ml 10 x TBE Pour as per demonstration

Sterile water	100ml bottle	
STOP soln.	50µl	0.25M EDTA pH8.0
T-mix	5µl	25µl 0.5mM dTTP 500µl 0.5mM dCTP 500µl 0.5mM dGTP 50µl 10mM ddTTP 1000µl TE buffer
TE buffer	10ml	10mM Tris-HCl pH 7.5 1mM EDTA
TM buffer	5µl	100mM Tris-HCl pH8.5 50mM $MgCl_2$
Universal primer soln.	5µl	
X-phosphate soln. (BOEHRINGER DIG vial 10)	50µl	
X-ray cassette	2 /class	
Gilson yellow tips (P20/P200)	1 pack	

Appendix V

Additional Requirements

Preparation of M13 bacteriophage

Requirements	Amount Required	Recipe
10ml sterile Pipettes	1 case	
30ml culture tube	1	
Eppendorf tubes	1 jar	
Gilson Pipetman P20	1	
Hershey's buffer	10ml	
Incubator (@ 37°C)	1 /class	
L-broth	20ml	10g Tryptone 5g yeast extract 5g NaCl 1g glucose

Microcentrifuge	1 /3 students
O/N culture of JM109	2ml
Pipette disposal jar	1
Plate of M13 plaques	1
Shaking incubator	1 /class
Soft L-agar	10ml
Gilson yellow tips (p20/P200)	1 pack

M13 RF DNA isolation

Requirements	Amount Required	Recipe
1ml sterile pipettes	1 case	
250ml centrifuge buckets	1	
Chloramphenicol	2ml	7mg/ml in 50% ethanol
Gilson Pipetman P20	1	
L-broth	150ml in a 500ml conical flask	10g Tryptone 5g yeast extract 5g NaCl 1g glucose
O/N culture of JM109	2ml	
Pipette disposal jar	1	
Shaking incubator	1 /class	
Sorval RC5B centrifuge	1 /class	
Stock of M13 bacteriophage		
10ml (@ 10^{10} pfu/ml)		
U.V. spectrophotometer + cells	1 /clas	
Gilson yellow tips (p20/P200)	1 pack	

Phagemid ssDNA isolation

Requirements	Amount Required	Recipe
1ml sterile pipettes	1 case	
30ml culture tube	1	
Gilson Pipetman P20	1	
Kanamycin soln.	10ml	$70\mu g/ml$
L-broth	10ml	10g Tryptone 5g yeast extract 5g NaCl 1g glucose
Shaking incubator	1	
Stock of bacteriophage M13 KO7	10ml (@ 10^{10} pfu/ml)	
Vitamin B₁ soln.	10ml	0.001% solution
Gilson yellow tips (p20/P200)	1 pack	

Plasmid DNA transformation

Requirements	Amount Required	Recipe
2ml sterile pipettes	1 case	
30ml culture tube	1	
Eppendorf tubes	1 jar	
Gilson Pipetman P200	1	
Glass spreader in beaker of ethanol	1	
Hershey's buffer	1ml	
Incubator	1 /class	
L-broth	2ml	10g Tryptone 5g yeast extract 5g NaCl 1g glucose
Microcentrifuge	1 /3 students	
Pipette disposal jar	1	
Shaking incubator	1 /class	
Gilson yellow tips (p20/P200)	1 pack	

Phosphotase treatment of vector DNA

Requirements	Amount Required	Recipe
37°C water bath	1 /class	
5 x loading buffer	5μl	50% glycerol 100mM EDTA pH8.0 0.125% bromophenol blue 0.125% xylene cyanol FF 0.25ml H$_2$O
Agarose	1g	
Alkaline phosphotase	2 units	
Chloroform	200μl	
CORE buffer	2μl	500mM Tris-HCl pH7.5 500mM NaCl 100mM MgCl$_2$ 10mM DTT
Electrophoresis buffer	500ml	30g glycine 1.2g NaOH 1.5g EDTA 2l H$_2$O 250μl Ethidium bromide solution

Eppendorf tubes	1 jar
Ethanol (@ -20°C)	100ml /class
Gel box and combs	1
Gilson Pipetman P20	1
Microcentrifuge	1 /3 students
Phenol/TE	200μl
Restriction enzyme	15 units
Sterile water	10μl
Gilson yellow tips (p20/P200)	1 pack
Vector DNA (e.g. M13mp18)	1.5μg

Preparation of sequencing gels (Non-gradient gels).

Requirements	Amount Required	Recipe
1 x TBE	1l	1:10 dilution of 10 x TBE
10 x TBE	15ml	109g Tris 55g boric acid 9.3g EDTA 1 litre H_2O
10% ammonium persulphate soln.	1ml	
250ml conical flask	1	
40% acrylamide soln.	25ml	38g acrylamide 2g Bis acrylamide 100ml H_2O
BRL sequencing apparatus	1 /class	
Measuring cylinder	1	
Power pack	1 /class	
TEMED	50μl	
Urea	50g	

Preparation of sequencing gels (Gradient gels).

Requirements	Amount Required	Recipe
1 x TBE	500ml	
1 x TBE gel mix	100ml	10.9g Tris 5.5g boric acid 0.93g EDTA 1 litre H$_2$O
10% ammonium persulphate soln.	1ml	
2 x TBE	500ml	1:5 dilution of 10 x TBE
25ml pipette	1	
5 x TBE gel mix	20ml	
BRL sequencing apparatus	1 /class	
Pi-pump	1	
Power pack	1 /class	
TEMED	50μl	

Fixing gels

Requirements	Amount Required	Recipe
Acetic acid soln.	1l /class	10% solution
Gel dryer	1 /class	
Glass rod	1 /class	
Large spactula	1 /class	
Large tray	1 /class	
Saran wrap	1 roll /class	
Scalpel	1 /class	
Vacuum pump + nitrogen trap	1 /class	
Whatman 3MM	1 sheet /class	
X-ray cassette box	1 /class	
X-ray developer etc.	1 dark room /class	

Appendix VI

Materials available for teaching practicals

A number of individual packages are available to people who have attended the DNA Cloning/Sequencing Course. These packages ar designed to allow the development of molecular biology practical classes based around the one week course taught as part of the BSc Molecular Biology at Portsmouth Polytechnic. They would particularly suit people who are introducing a new molecular biology course to their students.

Tick boxes for required items

1. **DNA Cloning Package**   £55.00

    DNA, enzymes, buffers and bacterial cultures required to complete M13 cloning.

2. **DNA Sequencing package**   £30.00

    DNA, enzymes, buffers required for DNA sequencing. Does NOT include radioactive material.

3. **Southern Blotting package**   £25.00

    DNA, enzymes, buffers and other materials required for Southern blotting. Does NOT include the Non-Radioactive Labelling kit.

4. **Plasmid isolation package**   £40.00

    DNA, enzymes , buffers required for the various plasmid isolation techniques. Does NOT include the column purification kit.

All of the packages include additional written material explaining what the expected results should be and the theory behind he materials used. They all require a variety of equipment (e.g. electrophoresis etc.) that can be obtained from the sponsoring companies. details are included with the packages.

Please tick the boxes above and send a photocopy to:

Dr. Keith Firman
DNA Cloning/Sequencing Packages
Biophysics Laboratories
Portsmouth Polytechnic
St. Michael's Building
White Swan Road
Portsmouth PO1 2DT.